George Thomas Stevens

Life as a Physical Phenomenon

Read before the Albany Institute, May 5, 1869

George Thomas Stevens

Life as a Physical Phenomenon
Read before the Albany Institute, May 5, 1869

ISBN/EAN: 9783337816469

Printed in Europe, USA, Canada, Australia, Japan

Cover: Foto ©berggeist007 / pixelio.de

More available books at **www.hansebooks.com**

LIFE

AS A

PHYSICAL PHENOMENON.

READ BEFORE

THE ALBANY INSTITUTE,

MAY 5, 1869.

BY

GEORGE T. STEVENS, M. D.

ALBANY:

J. MUNSELL, 82 STATE STREET.

1869.

Life as a Physical Phenomenon.

Naturalists have recognized as the first step in the classification of those beings of which we have knowledge, three great classes or kingdoms. These kingdoms are : mineral, vegetable, animal. Within these great classes are included all objects which come within the limits of human observation, whether visible to the unassisted eye or only revealed to vision by the aid of the most powerful glasses of the microscope; whether ponderous enough to be felt or so light and infinitely diffused that they float about in the atmosphere, a single volume of which outweighs a score of volumes of these subtle elements.

Curiously enough, the elementary constituents of beings in these three great divisions are in one sense common ; that is, all the elements of vegetables are also found in the class of minerals, and the elements of animal structures are none other than those of vegetables. Indeed, certain mineral elements, under certain influences, become plants ; and the tissues of plants under certain other influences, or possibly under the same influences, differently applied, assume the character of animals.

Individuals, which are extreme types of each of these classes, appear to indicate a most unmistakable line of demarkation between the classes ; no one could fail to see the distinction between a quadruped and a mushroom, or between a rock and a rose; but as we descend to the lower orders of vegetable and animal existence, it is often a question requiring the nicest discrimination to determine to which of the classes a being may belong.

There are even naturalists of profound learning and research, who declare that certain beings belong to both of these two classes, and are at one period of their existence plants and at another animals. This is mentioned not with a view of endorsing the theory, but to illustrate the extreme difficulty of separating the two classes. It is scarcely less difficult to draw the line of distinction between minerals and vegetables.

Notwithstanding the striking peculiarities pertaining to each of the kingdoms, when we reach beings sufficiently developed to represent to any certain degree the class to which they belong, there are very many characteristics which are common beside the material which enters into their composition.

The first common characteristic which we observe, aside from material, is a tendency on the part of the ultimate particles or molecules of each of the classes to arrange themselves in certain definite, specific forms; and if the circumstances surrounding any being be favorable, it will invariably assume the form peculiar to its class.

Any variation from this law is the result of opposing influences which the molecules cannot overcome. And as here is one of the first fundamental structural laws common to all beings, so here we observe the first indication of a line of distinction upon which any classification can be founded.

In minerals, where a specific form is assumed, the molecules unite in crystals, they are usually bounded by straight lines and angles, while vegetable and animal structures are built of cells; the prevailing idea is a sphere, and the parts are bounded by curved lines and surfaces.

Here, it would seem, that there is an absolute line of separation were it not for some exceptions which we may have occasion to mention.

If there is not an absolute line of classification in the general outline of minerals, as compared with plants and animals, we have but to advance one step and we find what seems to be a remarkable and invariable difference.

Plants and animals are always built up from cells, minerals, never.

Here then is our first absolute line of departure from a common type in the construction of beings, and it may be regarded as extremely doubtful whether we find another. Between plants and animals, it is doubtful whether there is more than the single distinction, a plant deoxydizes, an animal oxydizes.

Beings of neither of these classes are, where the typical form exists, formed by simple accumulation of particles. The assumption in most physiological works that minerals are built by simple accretion of particles is not true; but there is a process, wonderful in its perfection, by which certain particles, under certain circumstances, are arranged in the form of a crystal, and certain other particles, under certain other circumstances are made to constitute a cell. That the cell differs widely from the crystal all must admit; but that there is less perfection in the arrangement of the particles of the crystal than of the cell has never been demonstrated.

That peculiar formative process by which particles or molecules are arranged into cells has received various names, which are perhaps well enough in themselves but they have been so long used in a vague and undefined manner that the idea conveyed is often, if not generally, erroneous.

The phenomena which distinguish plants and animals from minerals are called collectively, *life*, or the vital process — vitality—by which terms is usually understood far more than seems to be indicated by the actual processes. Physiologists, recognizing the difficulties attending the use of words which seem to imply so much that is unknown and so little that is known, have substituted the word *organization* as the dividing characteristic between the first and the other two classes. Possibly this might convey a more definite idea if it could be assumed that there is no such thing as organization in minerals.

As I believe that it can be shown that the process of arranging the molecules of a crystal is as truly a process of organization as the arrangement of the molecules of a cell, and inasmuch as the terms *organization, formative process* should indicate generally and not specially, the process by which beings are formed, it would seem that there is little of definite meaning to the terms as used, and that so far as any idea is conveyed it is essentially erroneous.

If it is true that crystallization is really organization, it follows that we have two forms or varieties to study, the one working silently among minerals, the other building up the wonderful structures which we call living beings. Except in this cell form of organization there is no animation. Without this peculiar process, the earth would be a cheerless desert. There would be no longer green hillsides with flocks and herds; forests and groves, fields of waving grain and gardens of flowers would disappear; the singing of birds and the voices of childhood and youth would be hushed, and the gloomy earth would become a realm of desolation and death.

What is this process that builds the tree and paints the flower; that forms the beast that grazes upon the hillsides, and the bird that wings its way through the air; that gives the brilliancy to the eye and the bloom to the cheek?

It is impossible to begin the study of the laws which divide and unite the three great kingdoms in beings of complex nature; for so curiously and so mysteriously are the forces brought to bear, that we are confounded and overwhelmed by the surprising manifestations that meet us at the very outset of our investigations.

He who would learn to read does not begin at once to discuss and analyze the literary beauties of Milton or Shakespeare, but he learns first the simplest symbols and combinations. So when we would learn to read from the book of nature the secrets of *life* we are not to commence, as many do, with the most complex forms of existence; but we are first to learn what we can from the simplest forms, and trace the processes we find higher and higher as we become more and more familiar with the elements of truth they teach us.

The simplest living structure is a single cell. Some plants and some animals are composed of just this one cell. Now what can we learn from one of these minims of creation?

We find the green scum that floats on a stagnant pool, we place a little of it under a powerful glass, and we see that the scum is made up of hundreds of individual plants, each plant a single cell.

How does it look under the glass? It looks like a bag containing a transparent sac, within which is a fluid, and in the centre a dot; around the dot are many minute points or smaller dots arranged like a system of planets about a central globe.

Every one of these little plants of this simple construction consists of this outer bag, this inner sac, this contained fluid, and the dot. These cells or individual plants are very minute; a hundred of them would rest upon a pin's head.

Here then under the glass of the microscope we have a living being in which we can inspect each of its parts, and see all its little plans of life. Are these plans in any way similar to the operations in higher structures? Yes; what transpires within the limits of this little sphere is a type of all the operations carried on in every tissue in every living creature?

Indeed, at one period of its existence, every living thing, from the lowest plant to the highest animal, to man himself, was nothing more then one of these little spheres; and the most complex being can be traced back to its condition as a sac with its contained fluid and dot.

The higher forms of existence are made up by the multiplication in numbers and modification in form of these same cells ; and life in man himself is but the multiplication of what may be seen in this little cell under the glass.

Now let us see what goes on in the little being. If it continues to live under the glass we shall presently see that the minute granules, which are within the cell, are moving about. Watch them closely, and we find that they are all revolving about the nucleus or dot. We may observe what seems at first some confusion, for while the granules seem nearly all to be revolving in the same direction some very small ones seem to have an irregular motion; but by closer inspection we see that these smallest granules are revolving like satellites about other granules, which are larger and which have entered upon a stage of their existence hereafter to be described.

I shall now crave your indulgence for a moment while I call your attention to what seems to me at least a very curious analogy. I would like to leave the little cell with its revolving granules, and exchange the microscope for the telescope. If on some bright evening we go to yonder observatory, and watch the eye-piece of the telescope as the instrument is made to point to the stars, we shall gaze upon objects of wondrous interest. Jupiter and his satellites, Saturn with his moons and rings, and other planets are brought to view with greatly magnified splendor. As the instrument sweeps from one point of the heavens to another it brings to our notice a hazy, cloudy mass larger than hundreds of stars, looking in the distant heavens like an illuminated mist or as though some comet in its rapid flight had left a portion of its train behind. Such luminous masses are always found in the same part of the heavens, as fixed as are the fixed stars themselves. Astronomers call them nebulae

There is a theory now accepted by astronomers that our solar system was formed from just such a nebulous mass. Let me recall very briefly the main points of this theory.

The nebula of which our system was formed was set in motion by the power of omnipotence, and made to revolve. By its swift motion and by the attraction of particles for each other a central globe was drawn together, leaving the nebula divided into this central sphere, and a great ring of nebulous matter revolving about it. This central mass is the sun. At length the great ring, by the velocity of its motion, was broken into lesser rings, one within the other, all revolving in the same direction. The force of attraction and the swift revolution of the rings caused them again to break up, but not as before. Now instead of dividing into thinner rings the matter gathers into globular masses, and these masses revolve upon themselves while they continue to fly in the great orbit originally occupied by three several rings.

Here, then, we have a great central globe, around which was a ring of nebulous matter, which ring has broken up into smaller globes which revolve about the central one; and all these are surrounded by and contained in an atmosphere or ether, which, while it may be unlike our own, is still a surrounding and enveloping element for them all.

Now do we see between this system of globes and this little cell with its nucleus and revolving granules any resemblance? Is there not at least a very strong suggestion : that a single thought prevailed in the creation of the microscopic plant and the almost infinite system of worlds? And when we remember that this same principle of the

2

cell reaches through all phases of living existence (as I shall presently show it does) does not the suggestion become a conviction that a single mind has designed and created them all ; and that they are all, solar systems and living beings, designed on the same plan — different manifestations of the same grand idea, a single law governing them all ?

The little granules contained within the cell walls at length enlarge. They appropriate the nourishment in the little sac of fluid, and the sac in turn continues to draw more from the surrounding medium. The little granule is seen to collect about itself a thin transparent ring of matter which becomes an envelope like the original sac, and we are reminded of Saturn and his rings. Here, in one part of the field of the microscope is the miniature Saturn. There is another granule growing rapidly, around which other smaller granules revolve, reminding us of Jupiter and his moons. In another part of the field a little group of granules revolve together, recalling to mind some constellation, while the clear liquid in which this little cosmos is contained is like the celestial spaces. We cannot watch the process under the microscope without the conviction that the little plant is a repetition, on a small scale, of the systems of stars about us.

It may be said that the existence of cell walls destroys the analogy. That in one case we have but the circulation of a fluid within a closed cavity, while in the other we have the revolution of great masses in open spaces. But the cell wall is not essential, for these revolutions of granules may be seen in masses of protoplasm in which there is no indication of a cell wall, the granules are gathered about the nucleus and independently of any boundary but the attraction by which they are related to the nucleus, perform their revolution. Thus the analogy becomes perfect, the cell wall is an addition to the ideal cell, placed about it by infinite wisdom to protect the little system, and to assist in its relations with surrounding systems. Divesting the cell of its investments, and there remains but the difference in magnitude between the galaxy of the planets and the group of microscopic granules.

As the sac of the cell continues to draw within itself materials for protoplasm, and as the granules increase in size, they become too large to be contained in the parent cell, which bursts, and we see that each larger granule has become a new nucleus, has surrounded itself with cell walls like its parent, and is prepared for a separate existence. Each of the new cells is seen to teem with new granules; and each, as these granules enlarge, bursts and gives rise to a new brood of living beings.

Particles have been drawn from the mineral world, carbonic acid, water, and ammonia, have been taken up and decomposed, and certain quantities of oxygen returned to the atmosphere. Thus far, the process is purely chemical. But these materials are no longer carbonic acid, water, and ammonia. Each of these minims of the vegetable world has not only imbibed from the mineral kingdom certain elements, but it has assimilated this pabulum, converting it into products for its own enlargement or growth, and for the nourishment of still another class of beings. The elements which have been used in this process have taken a form never known to them in any other relations than when combined in living structures, and instead of remaining amorphous or assuming a crystalline form are arranged in cells. Whether the force which causes these particles when thus brought together in these proportions and under these peculiar circumstances is an inherent property of matter and corelated to the other forces which are manifested in the mineral world, or whether it is a new and peculiar force superadded to beings of this construction, and which is called *vitality*, is a question which we will examine when we have briefly considered how these units of living structures are related and combined in higher organisms.

We have seen the life process in the most simple structure, let us trace its workings in more complex beings. Cells do not always retain their spherical form, even when they are originally spheres. One of the first modifications is an elongation in the direction of growth.

Let us now advance one step in the process of existence.

Here I represent a being, a plant consisting of several such cells as have been described, and as we see, they are elongated in the axis of growth. The cells are connected at the extremities, and are thus united to make up a single individual. Now the little green filaments which we have all seen floating on the still waters, are, many of them, made up just in this way—a single row of cells, connected at the extremities.

Let us examine another form. Here is a filament made up of a single row of cells, but instead of being elongated they are flattened. This is the appearance of certain sea-weeds when placed under the lenses of the microscope.

Now, let us take a very small thread of muscular fibre, and place it in the field of the microscope. It will present the appearance of several of these little threads of sea-weed placed side by side, and if we separate these rows of cells and place a single ultimate fibre of muscle under the glass, it may be hard to tell which is the fibre of muscle and which the thread of sea-weed. Thus we see that a very large and important portion of the human body is built on the identical plan of the structure of the humble weed. And do we not see also how rapidly we have advanced in the scale of being? We commenced with the simplest living structure, the slime in the pool, then we examined a being scarcely higher in the scale of life, and then by the most natural and easy advance came directly

upon the principal tissue of man himself. Thus far we have
spoken only of cells that retain very nearly the typical,
spherical forms ; but many cells are so modified that a casual
examination detects no relation to the type. The cell is
modified in form according to the purpose it is to serve in
the economy of the individual. In the fibre of the linden
or oak it is long and flexible, more like a thread than what
we have figured as a cell. In what are called fibrous tissues too in the bodies of animals it is often drawn out in
the form of a long cylinder. Perhaps we may be able to
trace the transition from the round cell to the cylindrical
one so that we may keep the relationship in view.

Let us make a hollow sphere of india-rubber in imi-
tation of a round cell, the rubber taking the place of
cellulose. Let the sphere be large enough to illustrate the
changes we desire to witness and the walls thin enough to
be transparent. In the centre is a fluid, and within it
some elastic substance which may represent the dot. Now
let us fasten two fine hooks, one to one side and one to
the other of our elastic, transparent sphere. Draw upon
the hooks and do we not see that the rounded body becomes pointed ? Draw a little more, it is fusiform ; still
more, it is like a fibre, and now we can only see it as a
tube. The first, or rounded form is found in blood cor-
puscles in our own vessels, and discs of essentially the
same structure are also found in the mucous membrane lin-
ing the mouth, in the skin and in other tissues of the body,
and we have already seen how slight is its modification in
muscular fibre. The other forms are found in fibrous tissues. The long tubular form is common in woody fibres.
The fibre of cotton is an example of a cell greatly extended.

It might be shown, had we the time and were it necessary, that every part of a plant or an animal is built of
cells, but the proposition that such is the case will doubt-
less be accepted by all.

As I have already intimated, the cells draw from the surrounding fluids the materials necessary for their growth, and it is a remarkable fact that every kind of cell has the power, or faculty, of selecting and appropriating just such materials as are needful for its own peculiar structure. This faculty is, however, confined within certain limits. Thus the cells of plants will absorb carbonic acid, water, and ammonia, and unite these elements in the form of cellulose, but to the animal these materials would furnish only starvation diet. But these same compounds, carbonic acid water and ammonia, when once united in the plant, become food for the animal.

Having traced the history of the cell and its relations to living structures, thus far, it requires no stretch of imagination to perceive clearly that all living forms are funda-mentally alike, and that the cell or more exactly the protoplasm, from which the cell is built, is the real basis of all life. It is also easy to show that the movements of living beings are the result of changes of form in these structural units. Thus muscular contractility, which is regarded as one of the crowning marks of distinction between animals and all other classes of beings, results, simply from the flattening or shortening of multitudes of individual cells, and this effect may be produced by the action of electricity as well as by the force which passes along the line of a nerve, and this phenomenon of contractility is, by no means, confined to animals, but is found also in plants, and it might be shown that the identical principle exists also in minerals. But aside from this last statement, it is evident from what we have already seen, that the difference in faculty between the lowest plant or the lowest animal and the highest of its class is but one of degree, and not of kind. There would seem to be the widest contrast between the simple being we have figured as the unit of the vegetable world, and the giant tree that withstands the storms of ages, or between the deer that flies over the plains with fleetness and grace, and one of the microscopic inhabitants of a drop of water.

But as we trace the faculties of each to its origin we find that the contrast consists in the fact that in the lowest organisms each part is capable of performing all the functions of the being while in the higher structures the functions are distributed to different organs. This is an apparent rather than a real distinction, and hence it follows from this and from all that we have seen of the plan of living beings, that so far as the essentials of life are concerned we may regard the microscopic plant as a representative living structure; and in this light let us compare it with other beings which are not usually regarded as organized, and let us, if we are able, see where the process of organization begins, and when it becomes necessary to call to the aid of the ordinary forces and properties of matter, that peculiar and mysterious something called vitality.

Let us inquire whether we shall accept the theory that an all-wise architect has brought together the molecules of a cell, placing them in definite positions and arranging them for a specific purpose, while the atoms of a crystal have fallen together by chance and are the result, not of growth but of aggregation.

Certain elements unite to form protoplasm. They are carbon, hydrogen, oxygen and nitrogen, but these elements refuse to combine directly to form this protoplasm demanding first a binary union among themselves. Carbon and oxygen unite as carbonic acid, hydrogen and oxygen as water, and nitrogen and hydrogen as ammonia.

By the action of light, heat and electricity, which are but names for certain vibratory motions of matter, these compound molecules are made to move upon each other in such a way that protoplasm, a substance more complex in its chemical composition than mineral structures, and at certain ranges of temperature more unstable than some of them, is formed.

On the other hand, from a solution containing oxide of calcium and carbonic acid two binaries instead of three, a selection of atoms takes place. The liquid solvent with all the substances it contains, other than the two we have mentioned, are rejected, while the compound atoms of lime and carbonic acid, are united as a solid body having greater perfection of outline than the cell, and equal chemical purity. Here we have a crystal of calcareous spar; there we have a cryptogamous plant; the one is made of calcium, carbon, and oxygen, the other of carbon, hydrogen oxygen, and nitrogen. Neither of these groups of elements combine spontaneously under any and all circumstances. Pass a current of carbonic acid over a quantity of quick lime and you have not a crystal. Mingle carbonic acid, water, and ammonia, and you have not a plant. Around the plant are certain elements in solution which by a continual process of selection are assimilated to itself. Neither the animal nor the crystal differ from it in this respect.

The undulations of light, of heat, or of electricity, singly or in unison, cause these compound atoms to oscillate in harmony with the moving force, and as the particles are thus agitated by this tremulous molecular motion, the crystal or the cell result according to the nature of the ultimate particles interacting upon each other; and it is more than probable that could we but measure the waves that move the atoms, and weigh the molecules which enter into the structures, we might mathematically deduce the cell or the crystal from the materials and the force.

In considering the relations of minerals with plants and animals, we may perhaps soonest arrive at the real nature of these relations by examining some of the reasons almost universally accepted for regarding these classes as having no relations at all.

The principal features which distinguish what are called organic from inorganic bodies are thus given by one of the latest as well as one of the best and most complete works on physiology : [1]

1st. Minerals grow by accretion, plants and animals by living processes.

2d. The form of organic bodies is determinate; but a determinate form is not essential to minerals.

3d. In size organic bodies are determinate, while minerals deviate from any given standard of bulk.

4th. They differ in chemical composition.

5th. They differ in structure.

6th. Minerals are more permanent than organic structures.

7th. Organic bodies are all derived from parents.

Here, then, we have the full array of causes for *vitality*, and our author comments at length upon each of these distinguishing characteristics.

But will they stand the test of a rigid examination, so far as establishing the necessity of a new force is concerned?

" Minerals " are said " to grow by accretion, plants and animals by living processes."

Let us examine a living structure, and see whether it has any comment upon this proposition.

There exist in more or less abundance in most plants, crystals which often occupy the centre of a cell. In the tissues at the base of an onion may be found myriads of the most beautiful crystals. Here we have a compound crystal, in the form of a Greek cross, perfect in its symmetrical beauty, with its angles and planes more exact than could have been formed by the most skillful human mechanism, composed of four prisms grouped in such geometri-

[1] *Marshall.*

cal perfection as to excel the most complete work of art. Are we to be told that this exquisite gem, found within the walls of this cell, results solely from accretion of matter, while the envelope in which it is contained is the result of organization or of vitality?

But the "form of organic bodies is determinate, while a determinate form in minerals is not essential."

Properly viewed this is an absurd proposition. Matter has two determinate methods of uniting; in the form of the crystal, and in the form of the cell. We might with equal propriety select out the crystals, and say that because the remainder of matter is not invariably arranged in cells, therefore a determinate form is not essential to living structures.

It is also said that organic bodies are determinate in size, and that minerals are not. Are not the diamond, the calc-spar, the emerald and the garnet as determinate in size as the pine, the oak, or the cedar?

"They differ in chemical composition. The compound chemical substances contained in or derived from organic bodies are not imitated in the laboratory of the chemist, and must result from some force not known to ordinary chemistry."

This might have been a plausible proposition before Wöhler produced the organic compound urea, from inorganic elements in the laboratory or before tartaric acid, oxalic acid and other vital compounds were made without the intervention of any vital process.

"They differ in structure." "Organic bodies are composed of different parts having specific relations to each other. They are also heterogeneous as compared with the homogeneous structures of inorganic bodies, and from this heterogeneous structure arises the tendency and necessity during the life of an organized body of undergoing ceaseless internal motions and constant changes, while minerals may remain unchanged for an indefinite period of time."

Of course it is admitted that the structure of a plant differs from the structure of a mineral. It seems to be a fixed law of nature that when the compound atoms of water, ammonia, and carbonic acid, are subjected to certain influences, are agitated by certain ethereal waves, these atoms assume a certain form; while the simple atoms of oxygen and hydrogen, agitated by the waves of an electric current, assume a different form. The product of the four elements may be a sea-weed; the product of the two, a snow flake. This transformation by certain ethereal undulations into molecular rearrangements, which are unlike, demands for its explanation no new force.

We can hardly be required to explain the ceaseless changes, which are observed in living structures by any new force if they can be produced from forces already known to exist. It is a well established rule in science to eliminate unknown quantities from the problem when known quantities can supply their places as elements of the problem. Protoplasm is but the arrangement of certain atoms which are peculiarly subject to molecular change, and this may be said of all nitrogenous compounds.

The decomposition of substances containing nitrogen often evolve a sudden and surprising evolution of force. Witness the explosion of gunpowder when once its atoms are relieved from their unstable relationship by the vibratory action of heat. Mineral substances are only stable within certain limits.

The snow flake which falls upon your hand, a crystal of almost unparalleled beauty, is instantly resolved into an amorphous condition, and from this it is presently transformed into thin vapor. Can permanency of structure be claimed as a barrier to the relations of minerals and plants?

It may be said that the sudden destruction of the snow flake, or the more startling decomposition of gunpowder are unlike the ceaseless transformations which from hour to hour and from year to year may be observed in a living being of the higher orders. The cases are however alike. and the fact that one is an instantaneous change while the other is a succession of changes does not destroy the analogy.

But are there no gradual decompositions in the mineral world that correspond not only in the result but in time with those of living structures?

These ceaseless internal changes are the manifestations of an evolution of force, and force is never evolved without change in the relations of particles. Every contraction of a muscle, every exercise of the mind implies evolution of force with corresponding destruction of living tissue ; that is, a breaking up of the relations of atoms with each other and resolving live matter back to its mineral elements. If much force is evolved much organized matter is resolved, and all the varied manifestations of life depend on the amount of force.

The laborer who employs muscular force requires a large amount of material to replace the tissues broken down in maintaining the necessary movements. The student who labors severely with his brain may find on the following morning that phosphates in large quantity are eliminated.

Let us turn to the mineral world, and see if similar internal changes accompany the evolution of force.

Let us pour into a cup a solution of sulphate of copper; place in the cup also a strip of zinc and one of copper. The solution will remain for an indefinite time unchanged. But connect these strips by means of a wire, at once force is evolved, and we call that force electricity ; and just in proportion to the evolution of this force is the decomposition of the solution. Now is it not plain that these ceaseless internal changes in living beings in no way differ from those observed in minerals and that they are the necessary physical results of evolution of force? And were it necessary to illustrate this point further we might refer to seeds which after remaining buried beyond the disturbing influences which cause germination, for hundreds of years, have, when brought to the surface, sprouted and grown as readily as those which ripened but the autumn before. If a seed may remain alive for a thousand years with no internal changes why may it not for six thousand? If it is not subjected to certain forces which disturb the relations of its particles it may be as permanent as the granite hills.

So the solution of copper, so long as the disturbing element is not introduced, is a permanent solution; but once stretch the wire between the plates and let force be evolved, and the destruction of the solution has commenced.

There remains but the difference of origin. Why it is that protoplasm is not organized except by the aid of a nucleus of like nature we know not. But there are very many things in nature of which we can give no account. Are we assisted by bringing a name to our rescue, and saying that it is *vitality?* This method of resolving hard problems in science has been too long in fashion, and should be discarded.

Why not frankly confess that we have not found out the mystery of generation, and if it is worth finding out apply ourselves to the solution of the mystery and not cover it with a name which is but a dream of the imagination and carries with it not the shadow of any scientific intrinsic value?

For my own part I see no manifestations in living organisms which are not better explained by calling them the properties of matter than by calling them the result of *vitality;* and we have never been able to explain any phenomenon in which the ordinary, recognized forces of nature have not been our only aids.

Why then satisfy ourselves with mythical and mystical words which shall deceive us with an appearance of wisdom, but whose only office must be to point us away from the true path of investigation?